FLORA OF TROPICAL EAST AFRICA

MONIMIACEAE

B. Verdcourt

Trees, shrubs or rarely climbers, often aromatic; leaves exstipulate, opposite or rarely alternate, simple, entire or toothed (the teeth sometimes glandular). Flowers small, hermaphrodite or unisexual, sometimes polygamous, often dioecious, regular or rarely irregular, arranged in axillary or terminal racemes, panicles or racemes of cymules. Perianth-lobes 4–∞, small, imbricate in 2–several whorls, rarely absent. Stamens free, 6–∞; anthers dehiscing introrsely or extrorsely by longitudinal slits, more rarely by valves. Ovary superior, of 1–numerous 1-locular carpels; style long or short or stigma sessile; ovules solitary in each carpel, usually anatropous, pendulous or erect. Staminodes sometimes present in the ♀ flowers. Fruit of 1–∞ separate drupes or achenes, sometimes enclosed by the calyx or in or on the fleshy receptacle. Seeds with fleshy endosperm.

A family of about 20–35 genera and 150–350 species according to delimitation, occurring in the tropics and subtropics of both the Old and New Worlds, mainly in the southern hemisphere. A single monotypic genus (*Xymalos*) occurs in Africa, where it is an important element of the vegetation.

Hutchinson (Families of Flowering Plants, ed. 2, 1: 138 (1959) & G.F.P. 1: 124 (1964)) places *Xymalos* in the *Trimeniaceae*, but I agree with Takhtajan and Airy Shaw that this is not correct, the general structure of the flowers, particularly the stamens, being distinctly different. The monocarpellary ovary would appear to be an unimportant reduction. The *Trimeniaceae* are otherwise Pacific in distribution.

XYMALOS

Baill. in Bull. Soc. Linn. Paris 1: 650 (1887)

Paxiodendron Engl., P.O.A. C: 182 (1895)

Shrubs or small trees, almost entirely glabrous save for the inflorescences. Leaves opposite or subopposite, subcoriaceous. Flowers unisexual in axillary, solitary or paired velvety pubescent racemes or panicles. Male flowers: perianth 4–6-lobed; lobes ovate or lanceolate; stamens 10–15; anthers subsessile; rudimentary ovary nil. Female flowers: perianth 3–5-lobed; ovary of 1 carpel, obovoid, glabrous, a ring of hairs around its base; ovule solitary, pendulous; stigma thick, discoid. Fruit ovoid, fleshy, 1-seeded.

X. monospora (*Harv.*) *Warb.* in E. & P. Pf. III. 6a: 53, fig. 21/A, B (1893); Oliv. in Hook., Ic. Pl. 25, t. 2444 (1896); Perkins & Gilg in E.P. IV. 101: 23, fig. 4/G–L (1901); Sim, For. Fl. Cape Col.: 288, t. 121 (1907); Bak. & Wright in F.T.A. 6(1): 169 (1909); Perkins in E.P. IV. 101, Nachtr.: 10, fig. 5 (1911); Engl., V.E. 3(1): 219 (1915); T.S.K.: 6 (1936); T.T.C.L.: 349 (1949); Wimbush, Cat. Kenya Timbers: 69 (1950); F.P.S. 1: 7 (1950); J. Léon. in F.C.B. 2: 402, t. 39 (1951); I.T.U., ed. 2: 232 (1952); K.T.S.: 306 (1961); F.F.N.R.: 56, fig. 12 (1962). Types: South Africa, Orange Free State, *Cooper* 1204 & Natal, *Cooper* 1251 (TCD, syn., K, isosyn.!)

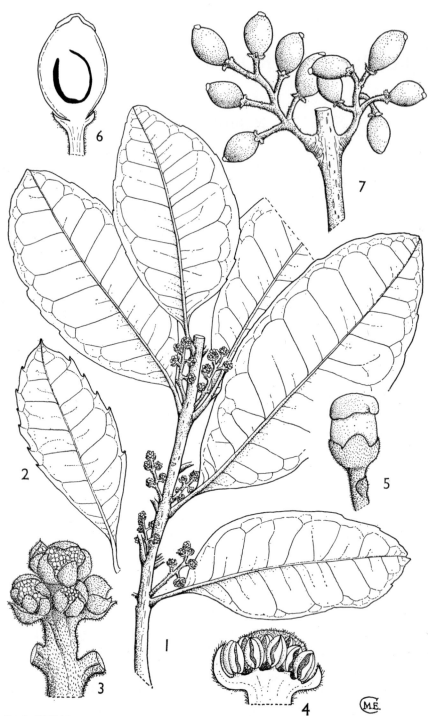

FIG. 1. *XYMALOS MONOSPORA*—**1**, part of male flowering branch, × ⅔; **2**, leaf, showing variation, × ⅔; **3**, part of male inflorescence, × 4; **4**, male flower in longitudinal section, × 10; **5**, female flower, × 6; **6**, female flower (late stage) in longitudinal section, × 6; **7**, part of fruiting branchlet, × 1. 1, from *Semsei* 2333; 2, from *Verdcourt* in *E.A.H.* 11862; 3, 4, from *Hughes* 70; 5, 6, from *Willan* 391; 7, from *Bally* 7604.

An evergreen shrub or small to medium-sized tree, 6–20(–27) m. tall, with glabrous branches; bark silvery-grey or grey-brown, rough, slightly fissured, with prominent lenticels, flaky, easily peeling; inner wood lemon-yellow; sap reddish. Leaf-lamina glabrous, lemon-scented, gland-dotted, narrowly elliptic to obovate, rounded, subacute or shortly acuminate at apex, cuneate at base, 4·5–20 cm. long and 1·8–10 cm. wide, usually ± irregularly and coarsely glandular serrate but sometimes entire; main lateral nerves 6–9, reticulate, prominent on both surfaces, looping well within the margin; petiole 0·5–2·2 cm. long. Flowers small, greenish, in velvety pubescent panicles or racemes (yellow-brown when dry), 1·0–3·5(–5) cm. long; peduncles 0·5–1 cm. long. Fruits orange or reddish, fleshy, ovoid, 1–1·9 cm. long, 1-seeded, crowned with the persistent stigma. Seeds ellipsoidal, compressed, 1 × 0·8 cm. Fig. 1.

UGANDA. Ankole District: Kalinzu Forest, June 1938, *Eggeling* 3657!; Kigezi District: Impenetrable Forest, Mar. 1947, *Purseglove* 2370!; Mengo District: Entebbe, June 1927, *Forest Dept. African staff* in *F.H.* 92!
KENYA. Northern Frontier Province: Mt. Kulal, July 1958, *Verdcourt* in *E.A.H.* 11862!; Meru, Aug. 1937, *V. G. van Someren* 242!; Teita District: Bura bluff, Chawia forest, 17 Sept. 1953, *Drummond & Hemsley* 4380!
TANGANYIKA. Bukoba District: Bushasha, *Gillman* 323!; N. Kilimanjaro, June 1934, *Pitt Schenkel* 363!; Lushoto District: Kitivo Forest Reserve, Aug. 1955, *Semsei* 2333!
DISTR. U1–4; K1, ?2 (West Suk), 3–7; T1–4, 6, 7; eastern Africa from the Sudan Republic and eastern Congo Republic to South Africa, also Cameroun Highlands and Fernando Po
HAB. Lowland and upland rain-forest, often a co-dominant in forests on isolated mountain-tops in dry country; 900–2700 m.

SYN. *Xylosma* (?) *monospora* Harv., Thes. Cap. 2: 52, t. 181 (1863)
    Toxicodendron acutifolium Benth. in J.L.S. 17: 214 (1878). Type: South Africa, " British Caffraria ", *Barber* (K, holo.!)
    Paxiodendron usambarense Engl., P.O.A. C: 182 (1895). Type: Tanganyika, E. Usambara Mts., Gonja, Bulwa, *Holst* 4249 (B, holo.†, K, iso.!)
    P. usambarense Engl. var. *serratifolia* Engl., P.O.A. C: 182 (1895). Types: Tanganyika, Kilimanjaro, Marangu, *Volkens* 1876 & 1908 & 2264 (all B, syn.†)
    P. ulugurense Engl. in E.J. 28: 389 (1900). Type: Tanganyika, Uluguru Mts., Lukwangule Plateau, *Goetze* 274 (B, holo.†, K, frag.!)
    Xymalos usambarensis (Engl.) Engl. in E.J. 30: 310 (1901)
    X. ulugurensis (Engl.) Engl. in E.J. 30: 310 (1901); Bak. & Wright in F.T.A. 6(1): 170 (1909); Perkins in E.P. IV. 101, Nachtr.: 11 (1911)

NOTE. *Xymalos monospora* yields an attractive and easily worked, durable timber, but it is little used.

INDEX TO MONIMIACEAE